MIX
The Chemistry Files

Discovery Channel School Science Collections

© 2000 by Discovery Communications, Inc. All rights reserved under International and Pan-American Copyright Conventions. No part of this book may be reproduced in any form or by any electronic or mechanical means, including information storage devices or systems, without prior written permission from the publisher. For information regarding permission, write to Discovery Channel School, 7700 Wisconsin Avenue, Bethesda, MD 20814. Printed in the USA ISBN: 1-58738-010-2

1 2 3 4 5 6 7 8 9 10 PO 06 05 04 03 02 00

Discovery Communications, Inc., produces high-quality television programming, interactive media, books, films, and consumer products. Discovery Networks, a division of Discovery Communication, Inc., operates and manages Discovery Channel, TLC, Animal Planet, Discovery Health Channel, and Travel Channel.

Writers: Stephen Currie, Bill Doyle, Vanessa Elder, Kathy Feeley, Scott Ingram, David Krasnow, Monique Peterson, Denise Vega, Anne Wright. **Editor**: Anne Wright. **Photographs**: Cover, pp. 2 and 12, swirl, ©PhotoDisc; pp. 2, 4, lemonade, Duncan Smith/Artville/PictureQuest; p. 4, salad, Duncan Smith/Artville/PictureQuest; girl eating BBQ, Timothy Shonnard/Stone; pp. 4–5, outdoor BBQ, Paul Redman/Stone; p. 6, rusty car, ©Nick Hawkes: Ecoscene/CORBIS; p. 8, dynamite bundle, ©PhotoDisc; p. 9, Kingdome imploding, ©Reuters Newmedia Inc./CORBIS; Nobel Prize medal, ©DCI/National Institutes of Health; p. 10, mixing bowl, © Digital Vision/PictureQuest; yeast under microscope, © David M. Phillips/Visuals Unlimited; apple, ©PhotoDisc; p. 11, bread dough, Painet Stock Photos; sliced onion, meat close-up, ©PhotoDisc (both); p. 13, statue damaged by acid rain, ©Alan Towse; Ecoscene/CORBIS; p. 16, Antoine Lavoisier, Brown Brothers, Ltd.; medieval alchemists, ©ArtToday; pp. 16–17, bucky ball model, Ken Eward/Science Source/Photo Researchers, Inc.; p. 18, Silly Putty, courtesy Binney & Smith Properties Inc.; p. 20, snail, ©CORBIS; p. 21, snake, ©Tom McHugh/Photo Researchers, Inc.; chameleons, ©Art Wolfe/Photo Researchers, Inc.; pp. 20–21, fireflies, Keith Kent/Science Photo Library/Photo Researchers, Inc.; p. 23, Ellen Swallow Richards portrait, courtesy MIT Museum; pp. 22–23, E. S. Richards taking water sample, Sophia Smith Collection/MIT Museum; pp. 24–25, Ran Tao taking photo, © Lori Berkowitz; p. 26, Sandra Steingraber portrait, ©Jerry Bauer/Courtesy of Jodi F. Solomon; pp. 26–27, field being sprayed with pesticides, ©CORBIS/Digital Stock; pp. 28–29, child on bike, ©PhotoDisc; p. 30, mummification, ©Bettmann/CORBIS; p. 31, chocolate, hydrangeas, ©PhotoDisc; all other images, ©Corel.
Acknowledgments: pp. 26–27, excerpts from LIVING DOWNSTREAM: A SCIENTIST'S PERSONAL INVESTIGATION OF CANCER AND THE ENVIRONMENT, by Sandra Steingraber. Copyright © 1998, Vintage Books, Random House. Reprinted with permission.

Stirring Things Up

When substances combine, cakes rise, clothes get clean, batteries power the radio—you can see all that without leaving your house. Now step outside: Grass grows, fires get extinguished, plastic is recycled, photos are taken— get the idea?

Chemistry is everywhere. It's the study of substances, their properties, and the ways they change when they combine. Peoples in the ancient world learned how to heat sand with other substances to create glass. Today, we can create fabric that's stronger than steel. We've also learned to make soap, aspirin, and plastics. The list could go on for pages.

But chemical processes don't always yield good results. Some chemicals pollute water, air, soil, and our bodies. Fortunately, we're learning from our mistakes, and doing more to prevent harmful chemicals from damaging the environment. Believe it or not, chemistry is often part of the solution. It is used to identify polluting substances and search for ways to counteract the damage.

In MIX, Discovery Channel takes you from junkyards to sophisticated labs to investigate the chemistry of the world around us. It opens up a whole new way of looking at things.

The Chemistry Files

Chemistry 4
At-A-Glance A backyard barbecue has its own special chemistry.

Spoilsport 6
Q & A Meet Rusty as he slowly destroys a car in a junkyard. Rust forms when three substances react, but you need more than a mixture of substances to get a reaction.

The Big Bang 8
Scrapbook For sheer drama, you can't beat a dynamite explosion. The fuse is lit and, within seconds, a tremendous force is unleashed, capable of leveling a building. Here's the inside story.

What's Cooking? 10
Picture This There's a lot of chemistry going on in the kitchen— when you cook, when you store food, and sometimes even when you're chopping vegetables.

The Acid Test 12
Almanac A substance's acidity gives chemists a great deal of information. The testing process can be quite colorful.

Practically Speaking 14
Map By accident or design, chemical discoveries have been improving daily life for thousands of years.

Thoroughly Modern Science 16
Timeline The science of chemistry has been around for hundreds of years, but there's always more to learn.

Battle of the Bounce 18
Scientist's Notebook Things don't always turn out the way scientists think they will. Searching for synthetic rubber, a chemist created Silly Putty™.

Curious Chemistry. 20
Amazing But True Chemistry plays a part in the lives of all animals. Some produce chemicals for defense, others to attract mates or capture prey.

Keeping Tabs 22
Eyewitness Account Ellen Swallow Richards knew that the chemistry of drinking water was crucial to public health.

What chemical helps this sky diver make a safe landing? Find out on page 17.

On a Roll 24
Virtual Voyage There's more to making a photograph than pointing and shooting. Follow film on its chemical journey from camera to finished print.

Crusader for a Safer World 26
Heroes A bout with cancer led Sandra Steingraber to look closely at industrial and agricultural chemicals in our environment.

Chemical Solution. 28
Solve-It-Yourself Mystery A bicycle has been stolen. All that's left is a broken lock—or is it? See how chemistry helps unravel the clues.

Chemistry Capers 30
Fun & Fantastic Chemistry makes for some strange combinations with even stranger results.

Final Project

Taking Stock 32
Your World, Your Turn No need to go to a lab to look for chemicals. Check out your kitchen, garage, or laundry room to find out what chemicals are on hand and how they work.

AT-A-GLANCE Chemistry

A scientist in a white coat mixes two liquids in a test tube. The mixture bubbles and spews—definitely a chemical reaction. But labs and test tubes aren't required for reactions to happen. They can happen anywhere—in the kitchen, in the washing machine, in the ocean, in hospitals, in our bodies. Chemists study reactions to learn what substances are made of and how they interact.

Every substance has both physical and chemical properties. Physical properties include texture, boiling and freezing points, taste, and appearance: Water freezes at 32°F (0°C) and boils at 212°F (100°C); lemon juice tastes sour. A chemical property tells us how one substance changes when it is mixed with another. If you mix vinegar and baking soda, carbon dioxide gas forms. You can see evidence of that gas in the bubbles that form.

But just combining substances doesn't always result in a chemical change. Spreading mustard on a hot dog changes the way the hot dog looks and tastes, but a new substance is not formed. Adding a drop of ink to a glass of water changes the appearance of the water and the ink, but does not form a new substance.

Most reactions need a little push to get started; they must be activated by energy, which can take the form of heat, light, or electricity. Energy plays other roles in chemical reactions. Some reactions release energy. A burning log is undergoing a reaction that releases heat. In other reactions, more energy is used than released. A plant uses sunlight in a reaction that produces the food it needs. Both chemical and physical changes are part of our everyday lives. You don't have to look any farther than your own backyard to find examples of both.

CHEMICAL

When we digest food, pepsin and hydrochloric acid in our stomachs go to work on our food, changing it chemically before it passes to other organs where the conversion process continues.

PHYSICAL

By adding sugar to lemonade, you change both the appearance of the sugar and the taste of the drink. You have made a solution—the sugar has broken down into tiny particles and mixed with the liquid, but it's still there and so is the lemonade.

PHYSICAL

Unless you shake it up, salad dressing has two separate layers—one oil and the other vinegar. Shake the bottle and the oil and vinegar form what's called an emulsion, a combination of liquids that don't normally mix. The oil breaks down into minuscule particles in the vinegar, but after a while, the oil particles get back together and the two liquids separate again—the lighter oil floats on top of the vinegar.

CHEMICAL

Burning charcoal is a chemical reaction in which the carbon in the charcoal and oxygen in the air react to produce carbon dioxide and water vapor.

As burgers grill, many complicated chemical reactions are taking place. The heat breaks down cell walls in the meat and the contents of the cells (minerals, amino acids, fats, carbohydrates, and enzymes) mix and react. The visible result of the reaction between the carbohydrates and amino acids is the brown crust on the burgers.

Spoil Sport

The life and times of a reaction called rust

Q: We have traveled today to our local junkyard to talk to rust. Hi there, Rusty. I must say, this is not where I'd choose to live. What brings you here?

A: Iron, water, and oxygen.

Q: I don't follow you. What do you mean?

A: I am the proud product of the chemical reaction of those three substances. You see, the car body that I'm attached to is made mostly of iron. The iron reacts with moist air (water and oxygen) to form me, hydrated ferric oxide. But you can call me Rusty.

Q: You say the iron reacts with the water and the oxygen. That sounds dangerous. Do you explode?

A: No, though I am super destructive, I'm not like those reactions that make a lot of noise or blow things to smithereens in seconds. I take my time and keep quiet. I do my work over days or even years. There is no time limit on a chemical reaction.

Q: Chemical reaction? Isn't that term a little grand for you?

A: Remember, chemical reactions can happen just about anywhere, even here—or should I say *especially* here. And I fit the reaction profile to a tee—I am a very visible product of the action of the three reactants I've already mentioned.

Q: I get it—like when you mix salt and water to make saltwater.

A: Hardly. That's only a physical—not a chemical—change. The salt is dissolved in the water so it looks different, but it's still there and so is the water. They haven't formed anything new. Just mixing substances together doesn't always create a chemical reaction.

Q: What if I throw in some heat, and, say, boil away the water? Don't tell me that's not a chemical reaction.

A: Sorry—it's still a physical change, nothing more. This time it's the water that has changed form—it has gone from being a liquid to a vapor, but it's still H_2O. And the salt—it's at the bottom of your pan, back in the form you started out with, more or less. Try again.

Q: How about flour and water—don't they form something new—paste—when they're mixed together?

A: Once again, chemically, nothing has happened. To prove it, all you have to do is put the flour and water mixture in a coffee filter and let it sit for a while. The flour and water will separate. It's hard to undo a chemical reaction—impossible in many cases.

Q: OK. I've got one—a burning candle. You can't undo that.

A: Bingo! Yes, where there's burning, there's a reaction. Some of the wax and all of the wick is burned, which produces gases, water, and ash. The wax that melts, though, has only undergone a physical change —it's still wax.

Q: Heat seems to be a key factor in the candle example. How does heat fit into your scenario?

A: What you mean is that heat is the energy source in that reaction. In your candle example, heat is also given off by the flame as the candle burns. Believe it or not, I give off heat too, but it's so subtle you probably wouldn't notice.

Q: Are there any other reactions out there like you?

A: No one can match me for wholesale destructiveness, but there are other corrosion reactions out there.

Q: I thought you said you were a chemical reaction. What's this about corrosion? Please make up your mind.

A: Let me explain. Corrosion is a type of chemical reaction in which a metal is worn away gradually by natural substances such as air and water, like the tarnish that forms on silver or that green stuff on the Statue of Liberty. Did you know there's copper underneath? But unlike me or my friend, Tarnish, that green stuff actually protects the copper because it seals it off from oxygen. It's called a vertigris.

Q: I do have some sense of your awesome power, but I must ask: Why do you not seem to have damaged the bumper of this car?

A: Ouch, that hurts! You are right. Bumpers are made of iron, but they are coated with chrome. And time and again, I am foiled by chrome, which prevents iron and oxygen (in the presence of water) from forming more of me.

Q: When I hear rust, I think reddish brown, but your color varies from a yellowish brown to black on different parts of this car. Please explain.

A: My color can change depending upon the amount of water present and the kind of metal I am working on. For example, if more water were introduced on the surface of this car—which could very well happen the next time it rains or snows—my color might change and become much darker, almost black, in places.

Q: How do you get around?

A: Once the metal has begun to rust quite nicely on one area of this car—note the beautiful work I have accomplished on the almost nonexistent trunk of this car—I can spread to other areas, as the reaction continues.

Q: Speaking of traveling, I need to be moving on. Thanks for talking to me today.

A: The pleasure was mine. I'm sure we'll meet again sometime—maybe on your car.

Activity

OXYGEN AT WORK Oxygen is a key component of many chemical reactions. Squeeze the juice from half a lemon into a small cup. Cut an apple in half. Put each half on a separate plate. Use a brush or your fingers to rub the cut surface of one apple half with lemon juice. Do nothing to the other half. Leave them out for an hour, then compare the two halves. Which half of the apple shows the characteristics of rust? What do you think kept the other half from "rusting"?

SCRAPBOOK: the BIG BANG

Chemical reactions that release heat are called exothermic. A burning log in the fireplace is an exothermic reaction: It gives off lots of heat. A dynamite explosion is exothermic, too. It releases a tremendous amount of energy in seconds, energy in the form of heat, light, sound, and wave motion in the air!

BLASTS FROM THE PAST

To dig through thick stone, builders in the ancient world, such as the Aztecs and the Romans, relied on a primitive technique. They heated the rock they needed to move, then cooled it quickly. The sudden temperature change was supposed to make the rock crack so the builders could break it apart. Whether it worked was a matter of chance.

About 1,000 years ago, the Chinese invented gunpowder by combining charcoal, sulfur, and potassium nitrate. In the 13th century, the Europeans made a similar discovery and began using gunpowder in weaponry and for blasting to clear building sites. But gunpowder is very volatile; the slightest shock can set it off, and there were many accidents. What's more, gunpowder blasts are not effective in breaking up rock.

Finally, in 1846, Italian scientist Ascanio Sobrero created nitroglycerin, an extremely volatile compound made up of nitrates and glycerol. It was powerful enough to be valuable as a blasting agent, but too dangerous to use. A minor tremor could cause an explosion.

Stockholm, Sweden, 1866
A BETTER WAY

Alfred Nobel, a Swedish chemist, found a way to preserve nitroglycerin's explosive power *and* decrease its volatility. First, Nobel mixed the nitroglycerin with non-volatile clay. The mixture was much harder to detonate but, once detonated, was still powerful. He packed the mixture into cardboard tubes and attached a blasting cap—a simple wooden plug filled with gunpowder with a fuse. Only by lighting the fuse and detonating the gunpowder could an explosive reaction be set off. The problem of unpredictable reactions was solved.

Dynamite was much safer than gunpowder or nitroglycerin alone, but there were occasional accidents. The nitroglycerin sometimes crystallized, meaning it was no longer mixed with the stabilizing clay. It could also sweat to the surface of the sticks or puddle in the crates in which it was stored.

AN INSIDE VIEW OF DYNAMITE

1. The blasting cap detonates, providing the initial shock that ignites a chemical reaction in the stick of dynamite.
2. The chemical reaction happens in a fraction of a second and travels from one end of the dynamite to the other.
3. The reaction produces a growing amount of hot gas, which forms a moving high-pressure zone.
4. The area in front of the chemical reaction is a low-pressure zone because the temperature stays the same. The change from low to high pressure happens so fast that a tremendous force is created.
5. This force can't be contained in the stick of dynamite. And an explosion occurs, blowing apart whatever is in its way.

Nobel Prize medal

PEACE OF MIND

Alfred Nobel was shocked to see himself referred to in a newspaper as the "Merchant of Death." To counteract this perception, he established the Nobel Prizes to be given for achievements in the sciences and literature and, most important, for peace.

AN EXPLOSION OF USES

The Panama Canal connects the Atlantic and Pacific Oceans. By the time it opened in 1914, it had been a dream for centuries. And dynamite made it happen. It took 67 million pounds (30.4 million kg) of dynamite to move 175 million cubic yards of earth (enough to fill the Empire State Building 500 times).

Dynamite has been used to prevent or minimize damage, too. Because a dynamite explosion decreases the amount of oxygen in the air, it helps contain fires, which need oxygen to burn. In 1991, during the Persian Gulf War, as Iraqi troops retreated from Kuwait, they set oil fields on fire. The fires raged out of control. American troops exploded dynamite near the wells, and the fires died down long enough for the troops to cover the wells.

To make way for a new stadium, builders detonated dynamite to bring down the Kingdome in Seattle, Washington.

The Heart of the Matter

Heart patients are sometimes treated with nitroglycerin. It makes blood vessels expand, enabling the blood to flow more freely and lowering blood pressure to help prevent heart attacks. The amount of nitroglycerin in a tablet is tiny and not harmful.

Activity

HEAT WAVE You wouldn't need a thermometer to tell you that a dynamite explosion gives off heat. What about reactions that aren't so dramatic? Try this: Put an outdoor thermometer in a clean jar with a lid. Wait five minutes. While you're waiting, soak a small steel wool pad (without soap) in 1/4 cup of vinegar for a minute or two. After the five minutes have passed, record the temperature in the jar, then remove the thermometer. Squeeze the steel wool as dry as you can (but don't rinse it) and wrap it around the base of the thermometer. Put the wrapped thermometer back in the jar and cover the jar with the lid. Wait five minutes, then take the temperature and record it. Unwrap the thermometer, rinse, and dry it. Using a clean, dry jar and a dry (unsoaked) piece of steel wool, repeat the procedure and record the temperature. What happened? Why? How would you characterize the reaction that took place?

THE CHEMISTRY FILES

What's Cooking?

Glad you asked—a lot of it is chemistry. Chemical and physical changes happen all the time in the kitchen. Some chemical changes, like dissolving yeast in a mixture of sugar and water, give off heat. Others take heat rather than release it. Take a look and see what's really going on while you're working in the kitchen.

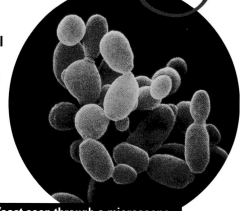
Yeast seen through a microscope

Better Baking

To make gingerbread, you start by mixing together flour, sugar, spices, and a little baking soda. The mixture contains separate ingredients, but no chemical reaction has yet occurred. When you stir in the liquid—molasses, buttermilk, or lemon juice—the chemical changes start. That little bit of baking soda reacts with the liquid to produce a light, airy cake.

Here's how it works: When the baking soda, a base, and the acidic liquid combine, the bitter taste of the baking soda is neutralized; carbon dioxide gas and heat are released, forming bubbles in the batter. (To learn more about acids and bases, turn to the Almanac, pages 12–13.) In the hot oven, the bubbles get bigger, making the batter rise. As the batter becomes cakelike, the bubbles can't expand any more so the carbon dioxide floats into the air, leaving behind a delicious dessert.

Fruit Facts

Compare the texture of a crisp apple to applesauce. Quite a difference when you consider that all you did was add a little water and turn on the heat. The heat breaks down the raw fruit's cell structure and releases water. The substances holding the cells together are also affected by the heat. They are converted to pectin and combine with the water. The result is smooth applesauce.

Don't Make Me Cry

If you don't like onions, it may be because they make you cry. An onion contains two substances, sulfoxides and allinase; when they're combined, they cause tears. In a whole onion, the two substances are kept separate by cell walls, but the physical change of slicing the onion cuts through the walls and allows the two substances to mix and form a new substance. Some of the new substance floats up to your eyes and reacts with the tears. The combination of tears and the new substance forms an acid. That's what stings your eyes and makes you cry.

Put Freezer Burn on Ice

What are those leathery white blotches you sometimes see on frozen meat? Some are ice that forms when water in the meat freezes—a physical change. But the greenish-gray crystals tell a different story. Freezing concentrates many of the chemical ingredients in food at the surface. As they react with each other and air, they can cause this unpleasant result.

Rising to the Occasion

Bread needs air bubbles, too. And it's yeast that usually makes them happen. A one-celled fungus, yeast feeds on sugars in the bread dough, converting the sugar to carbon dioxide and alcohol. This conversion is a chemical reaction that makes bread rise. The carbon dioxide forms bubbles. As the bubbles grow, the other ingredients move to make room for them. This reaction releases heat, but heat also helps the process along. Yeast performs best in a warm environment, about 80°– 95°F. By the time the bread goes into the oven, the carbon dioxide has done its work, and the dough is nice and puffy. But what happens to the yeast? It dies at temperatures of about 140°F and higher. And the alcohol? It evaporates. All you're left with is a luscious loaf of homemade bread.

Activity

SHELL GAME There's more than one way to crack an egg shell. Ask your parents for two eggs. Place them in two clean one-pint or larger jars with lids. Be careful not to crack the eggs. Add enough cold water to one jar to cover the egg by about an inch. To the other jar, add the same amount of white vinegar. Place the lids on the jars. Do you notice anything happening? Write down what you see. Leave the jars in place for 24 hours, but take a look from time to time and write down what you observe. After 24 hours, compare the eggs. What differences do you observe? What do you think caused any differences? Note: Discard both eggs at the end of this experiment. They may contain harmful bacteria. Be sure to wash your hands thoroughly after touching the eggs.

The ACID Test

One thing chemists want to know when they investigate liquids (and some solids) is whether or not they're *acids* or *bases*. It's one factor they consider when trying to predict how one substance will react with another.

You've heard of acids. You might have even tasted a couple—lemon juice and vinegar. A sharp taste is characteristic of acids (though many acids are NOT for tasting). A base is the opposite of an acid, to chemists at least. Ammonia, baking soda, and many soaps are bases. If you tasted soap, it would be bitter (many bases are not for tasting either). If you rubbed some on your hands, it might feel filmy. Not all acids are exactly the same—some are strong and others are mild. The same is true for bases. And, there's one substance that's neither—distilled water. It's what is known as *neutral*.

On a Scale of 0 to 14

Scientists describe how acidic or basic a substance is by where it falls on the pH scale. The scale measures acidity. The scale goes from 0 (most acidic) to 14 (most basic) with distilled water right in the middle: It has a pH of 7. Each integer is 10 times the power of the previous one. Look at the chart and you'll see that most of the substances we come in contact with on a day-to-day basis fall somewhere in the middle of the scale. The substances at the extreme ends of the scale are very strong—they can cause severe burns.

When dipped into different liquids, this specially treated paper, called litmus paper, indicates acid content.

Substance	pH level	acid, base, or neutral
Sodium Hydroxide	14.0	base
Lye	13.0	base
Ammonia	11.0	base
Milk of Magnesia	10.5	base
Baking Soda	8.3	base
Human Blood	7.4	base
Distilled Water	7.0	neutral
Milk	6.6	acid
Molasses	5.5	acid
Tomatoes	4.5	acid
Dried Apricots	4.0	acid
Apples	3.0	acid
Vinegar	2.2	acid
Battery Acid	1.0	acid
Hydrochloric Acid	0	acid

Showing Their True Colors

One way to measure an acid's strength is with a paper strip that has been specially treated with chemicals. These chemicals react by changing color in a predictable way when they come into contact with various levels of concentration. Like the scale below, universal indicator paper uses a scale of 0 to 14 to show the acidity of different substances.

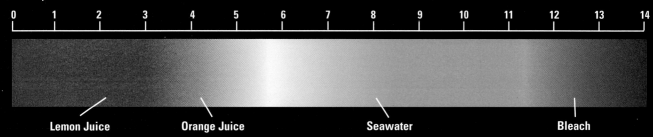

Lemon Juice — Orange Juice — Seawater — Bleach

By the Numbers

The abbreviation "pH" stands for the "potential hydrogen." Technically, the numbers on the pH scale refer to the concentration of hydrogen ions in the substance being tested. (Learn more about hydrogen ions on the Internet at www.Chem4Kids.com).

Mix-Ups

What happens to acidity levels when a base is combined with an acid? The acidity level decreases. A bee injects you with an acid when it stings. To counteract the stinging of the acid, you can rub the sting with a paste of baking soda (a base) and water, which will increase the pH level and relieve some of the discomfort caused by the acid.

Acid in the News

Stone sculpture damaged by acid rain

If you listen to the news or read the papers, you've heard about acid rain. The normal pH for unpolluted rain is 5.6, but acid rain has a pH of about 4.6. That means the acid content is about ten times higher than normal. The increased acidity is caused by pollution such as car exhaust and industrial emissions that have been released into the atmosphere over the years. The rain becomes acidic when the emissions in the air react with the rainwater. Acid is corrosive, so when acid rain hits a statue or a building, it starts to wear away the material. When acid rain hits the soil or a water source, it may dissolve important nutrients, making them unavailable to plants and fish that need them.

Activity

A PENNY SAVED Pennies are made of metals, which tarnish, or darken and lose their shine, when they come in contact with oxygen and moisture. Test which works best to clean a tarnished penny. In a cup, pour 2 tablespoons of vinegar. In another cup, combine 1 tablespoon dishwashing liquid and 3 tablespoons of water. In a third cup, place 4 tablespoons of water. Drop a tarnished penny into each. Wait 30 minutes, then take a look to see which penny is cleanest. What can you conclude about acids and tarnish on pennies?

Try this experiment with five other types of coins or metals. If you can, make one of them a foreign coin. Compile your results on a chart.

MAP Practically Speaking

For thousands of years, humans have been making chemical discoveries that make everyday life easier. Innovations in chemistry have brought us glass, bronze, aspirin, detergent, and synthetic fabrics. This map shows where a few of those discoveries were made.

❶ KEVLAR: Wilmington, Delaware Kevlar™ is a thin fiber five times stronger than steel. Dupont chemist Stephanie Kwolek produced the crystalline polymers, long chainlike molecules, from which Kevlar is made. Because it is strong, durable, weatherproof, flame-resistant, and lightweight, Kevlar is used in bulletproof vests, radial tires, spacecraft shells, skis, and camping gear.

❷ FIRE RETARDANT: Jupiter, Florida In October 1988 firefighter John Bartlett was battling a house fire when he noticed that a diaper hadn't burned. He found that the diaper contained an absorbent polymer gel that had fire-retardant qualities. Bartlett made a few modifications and produced a gel he called "Barricade." Firefighters sprayed it on 20 homes threatened with forest fires. None burned.

❸ QUININE: Bolivia and Peru In the 1600s, Spanish settlers reported that a powder made from the bark of the cinchona tree could cure malaria. The curing substance was a quinoline alkaloid, which scientists later named quinine.

❹ MATCHES: England Starting a fire was hard until English chemist John Walker created the friction match in 1826. He was stirring antimony sulfide, potassium chlorate, gum, and starch with a stick. When he took the stick out, the stuff dried. To get it off, he rubbed it on the stone floor. *Pffft!* The friction caused the chemicals to burn, and the friction match was born.

❺ SOAP: Belgium People have made soap for centuries, but until recently, it was considered a luxury. Soap was made from expensive materials. But in 1850, a Belgian chemist figured out how to make soap easily using inexpensive table salt. The result? Widespread manufacturing of cheaper, better soap.

❻ DETERGENT: Germany During World War I, Germany experienced a shortage of fats needed to make soap, so chemists developed a synthetic alternative—detergent, made up of alcohols and naphthalene, a chemical compound. Detergents did not produce the scum that occurs when soap combines with minerals in water. After additional refinements, detergent surpassed soap in cleaning power.

❼ ASPIRIN: Greece Hippocrates prescribed willow bark for headaches in about 400 BC. The bark contains salicylic acid, a distant relative of the active ingredient in aspirin. But it would be centuries before aspirin became widely available. More than 2,000 years later a German chemist made salicylic acid in a lab. How does it work? The salicylic acid reduces the body's sensitivity to pain.

❽ GLASS: Egypt Ancient Egyptians were probably the first to make glass. Using heat, they fused together sand (mainly silicon dioxide), lime (calcium carbonate), and soda ash (sodium carbonate). At first, they used glass for beads. Later, they used it to make bottles, goblets, bowls, and vases.

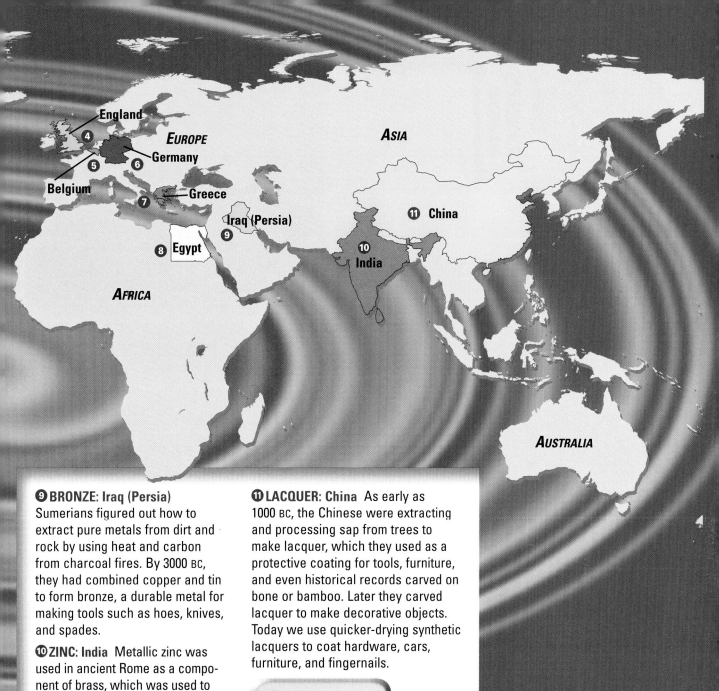

❾ BRONZE: Iraq (Persia)
Sumerians figured out how to extract pure metals from dirt and rock by using heat and carbon from charcoal fires. By 3000 BC, they had combined copper and tin to form bronze, a durable metal for making tools such as hoes, knives, and spades.

❿ ZINC: India Metallic zinc was used in ancient Rome as a component of brass, which was used to make armor and jewelry. But zinc was not isolated as a separate metal until the 13th century in India. Today zinc is used to coat iron and steel to prevent rusting, to make paint, cosmetics, and some types of batteries. And it is the main component of "copper" pennies. Researchers are studying zinc's medicinal properties to determine if it might cure the common cold.

⓫ LACQUER: China As early as 1000 BC, the Chinese were extracting and processing sap from trees to make lacquer, which they used as a protective coating for tools, furniture, and even historical records carved on bone or bamboo. Later they carved lacquer to make decorative objects. Today we use quicker-drying synthetic lacquers to coat hardware, cars, furniture, and fingernails.

Activity

KEEP CLEAN If you didn't have soap or detergent, what could you use to clean your favorite shirt? Do a test. Ask your parents for a piece of white cotton rag and spread it on a work surface. Make two grease spots on the cloth by rubbing it in two separate places with a little butter or margarine. Now squeeze a teaspoon or two of lemon juice on a paper towel and rub the paper towel into one of the stains. Then dip another paper towel in some whole milk and rub the other stain with the towel. Which substance cleaned the stains better? Try staining the cloth with other substances like cranberry juice, chocolate, and ink. Repeat the tests with the lemon juice and whole milk. If there is a different result from the margarine stains, can you think of reasons why this is so? Chart your results.

Timeline: Thoroughly Modern Science

Chemistry has been practiced since ancient times. In the ancient world, Egyptians made dyes; a little later Chinese made gunpowder using chemical processes.

Alchemy (AL-keh-me), the "science" of turning non-precious metals into gold, flourished first in the Arab world, then in Europe in the Middle Ages. The alchemists didn't achieve their goal, but some of their research methods and equipment proved valuable to scientists in later generations. Chemists also built on the work of early metallurgists, craftspeople who worked with metals—purifying, combining, and creating objects from them.

By the 17th century, Irish scientist Robert Boyle had learned much about the relationship of gases to pressure. And he showed that air, earth, fire, and water were not pure substances, but were made up of smaller pieces. Modern chemistry was just around the corner.

1780

French chemist Antoine Lavoisier shows that in a chemical reaction the total weight of the products always equals the original weight of the reactants. This rule is known as the Law of Conservation of Matter.

1800

Alessandro Volta builds the first battery, which shows that chemical reactions can produce electricity. The battery consists of layers of zinc disks and copper disks with saltwater-soaked cardboard in between. Later, scientists will discover that the electricity produced by the battery can be used to combine substances to produce new chemicals or break down complex substances.

1850

Engineers Henry Bessemer and William Kelly each develop an easy way to produce steel, which is derived from iron. In Bessemer's method, melted iron is poured into a special tank, then air is added. The oxygen in the air reacts with the impurities in the iron, making the impurities easy to remove.

Medieval alchemists

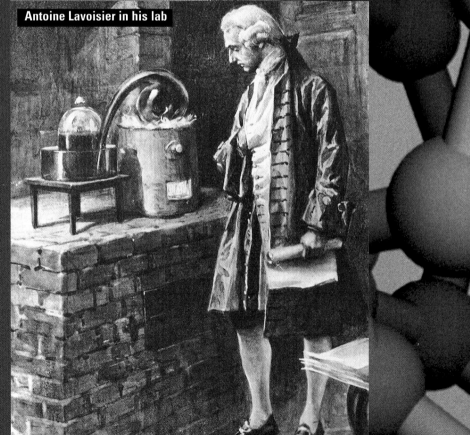

Antoine Lavoisier in his lab

16 DISCOVERY CHANNEL SCHOOL

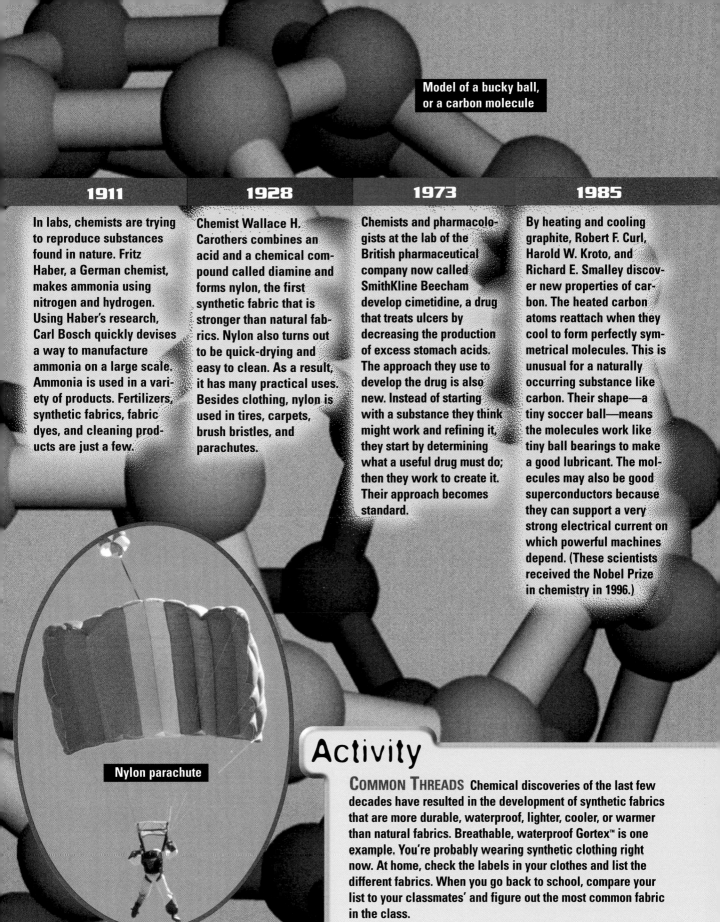

Model of a bucky ball, or a carbon molecule

1911

In labs, chemists are trying to reproduce substances found in nature. Fritz Haber, a German chemist, makes ammonia using nitrogen and hydrogen. Using Haber's research, Carl Bosch quickly devises a way to manufacture ammonia on a large scale. Ammonia is used in a variety of products. Fertilizers, synthetic fabrics, fabric dyes, and cleaning products are just a few.

1928

Chemist Wallace H. Carothers combines an acid and a chemical compound called diamine and forms nylon, the first synthetic fabric that is stronger than natural fabrics. Nylon also turns out to be quick-drying and easy to clean. As a result, it has many practical uses. Besides clothing, nylon is used in tires, carpets, brush bristles, and parachutes.

1973

Chemists and pharmacologists at the lab of the British pharmaceutical company now called SmithKline Beecham develop cimetidine, a drug that treats ulcers by decreasing the production of excess stomach acids. The approach they use to develop the drug is also new. Instead of starting with a substance they think might work and refining it, they start by determining what a useful drug must do; then they work to create it. Their approach becomes standard.

1985

By heating and cooling graphite, Robert F. Curl, Harold W. Kroto, and Richard E. Smalley discover new properties of carbon. The heated carbon atoms reattach when they cool to form perfectly symmetrical molecules. This is unusual for a naturally occurring substance like carbon. Their shape—a tiny soccer ball—means the molecules work like tiny ball bearings to make a good lubricant. The molecules may also be good superconductors because they can support a very strong electrical current on which powerful machines depend. (These scientists received the Nobel Prize in chemistry in 1996.)

Nylon parachute

Activity

COMMON THREADS Chemical discoveries of the last few decades have resulted in the development of synthetic fabrics that are more durable, waterproof, lighter, cooler, or warmer than natural fabrics. Breathable, waterproof Gortex™ is one example. You're probably wearing synthetic clothing right now. At home, check the labels in your clothes and list the different fabrics. When you go back to school, compare your list to your classmates' and figure out the most common fabric in the class.

SCIENTIST'S NOTEBOOK

Battle of the Bounce

New Haven, Connecticut, 1943

It was no laughing matter: It was World War II. And from the look on James Wright's face, he didn't have fun and games on his mind. In 1943, the United States and its allies were fighting against Nazi Germany and Japan. Underneath every jeep and plane was a complex substance known as rubber.

The Race for Rubber

Rubber comes from the sap of certain trees. When boiled and chemically treated, it hardens but remains flexible, snapping back into shape after being stretched. Most commercially produced rubber comes from areas of Asia that, during the war, were blocked by Japan. The search for a substitute became a national priority.

Enter James Wright. A native of Scotland, Wright was hard at work trying to re-create rubber in his lab by linking small molecules to form a long chain. This process is called polymerization. A polymer is a kind of super-molecule made up of thousands of small units that are linked together like a daisy chain. Rubber from trees is a natural polymer. (Plastics are synthetic polymers.) Because the polymer chain is so long, rubber can be stretched without breaking. Wright's challenge was to find another substance that would do the same thing.

Mistaken Identity

Wright thought silicon, an element found in sand, rocks, and soil, was a good candidate because its molecules formed chains like rubber molecules. And silicones, plastics made from silicon, can be molded, as can rubber. Wright tried mixing silicone oil with boric acid, a food preservative also used to kill cockroaches. He pulled the hardened substance out of the beaker, and . . .

It bounced. And when Wright squeezed the sample, its shape changed—it was not elastic at all. Interesting, but not the result he wanted. In fact, you could mold it into just about any shape and it would stay that way. Imagine landing a plane on wheels of putty. Tomorrow would be another long, long day at the lab.

It turned out that the secret

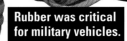

Rubber was critical for military vehicles.

Earning from Mistakes

SOMETIMES "FAILURES" PRODUCE USEFUL RESULTS.

Natural rubber is temperature sensitive. In hot weather it gets sticky; in cold weather it becomes brittle. In the winter of 1839, Charles Goodyear was looking for ways to make rubber more durable. He tried mixing it with different chemicals without luck. When he mixed the rubber with sulfur, he spilled some of the mixture on a hot stove by mistake. As he was cleaning up, Goodyear noticed that the heat from the stove did not make the mixture sticky. Instead, it was firm and flexible. He left some of the mixture outdoors overnight to see what would happen at cold temperatures. The texture didn't change. The heating process, which Goodyear called "vulcanization" after the Roman god of fire, had done the trick. Rubber became much more usable as a result.

In the 1850s, English chemist William Henry Perkin tried to make a synthetic quinine (used to treat malaria) by mixing salt of aniline with coal tar. The result was a black sludge at the bottom of the beaker. To clean it, he added alcohol and the mess turned a purple color. Perkin found that the mixture could dye silk. The concoction turned out to be the first synthetic dye.

of synthetic rubber was in petroleum. By breaking petroleum down into its parts, chemists found that certain molecules of carbon and hydrogen could be made to act like rubber. By 1945, American factories were producing one million tons of synthetic rubber a year.

A Use for Everything

After the war, Wright's employer, the General Electric company, sent samples of the bouncing putty to engineers. Useless, they declared. But eventually a sample found its way into the hands of Ruth Fallgatter, who owned a toy store in New Haven, Connecticut. With the help of marketer Peter Hodgson, she offered Silly Putty™ as a novelty item in her catalog.

People have been thinking up uses for Silly Putty™ ever since. Pressed against the colored newsprint of the comics, Silly Putty™ picks up a detailed image. The putty dissolves a tiny amount of the ink, and the silicone makes it stick without running. The astronauts on the Apollo 8 mission used it to hold tools in place. Some people squeeze it to relieve stress. Today, the makers of Silly Putty™ are proud of its quality that caused James Wright despair: "It's not like rubber, or anything else."

Chemist Spencer Silver worked for 3M, the company that makes Scotch™ tape. In 1970, he tried to develop a superstrong adhesive but came up with a superweak one instead. A surface coated with Silver's adhesive stuck, but the lightest tug would remove it. Later one of Silver's colleagues, Arthur Fry, remembered this weak adhesive. He sang in a choir and wanted the markers he used to keep his place in the hymn book to stay in. Fry coated the small squares of paper with Silver's weak adhesive, and the sticky note called the Post-it™ was born.

Activity

STICK TO IT Coming up with a good invention means a lot of testing! See what it's like. Get four different kinds of tape (such as Scotch tape, masking tape, duct tape, and a Post-it) and test them on various surfaces. Some examples might be a cloth rag, an old newspaper, a block of wood, and an ice cube. Observe what happens with each example on each surface, and make a chart of your findings. What specific characteristics can you describe about each type of adhesive? What are the best uses for each type?

THE CHEMISTRY FILES

CURIOUS CHEMISTRY

Biochemists study the chemistry of living things. For animals, chemistry is key—not only to their internal systems—but to the way some interact. Animals may produce and use chemicals in self-defense, to capture prey, for courtship, and to communicate. Here's how it works.

SEEING THE LIGHT

Some species of living things give off light, including several species of insects. But fireflies, also called lightning bugs, can flash their lights on and off.

To get that twinkle, a firefly's light-producing chemicals (luciferin and luciferase) underneath its abdomen react with oxygen in a process called bioluminescence. A firefly's light is energy efficient: It is almost all light, not heat. (About 90 percent of a household light bulb's energy is given off as heat.)

To attract a mate, a male firefly flashes his light in distinct patterns. If the female firefly finds his patterns attractive, she shines her own light in response. Drawn by the light, the two fireflies find each other.

ANIMAL MAGNETISM

Animals use pheromones, a class of chemicals secreted by their bodies, to send signals to other animals of the same species. Animals produce and release pheromones for many reasons: to establish territory, to warn others of danger, and to attract mates.

Certain mice, ants, and snails secrete alarm pheromones when they are injured or threatened, which alert the others. If a mouse meets with a cat, other mice will sense the alarm pheromone and stay away! The queen bee of a beehive releases a pheromone that prevents the other female bees from mating and laying eggs.

Scientists have developed artificial pheromones to control insect pests. Farmers use these pheromones to interrupt insect reproduction or to bait traps for crop-destroying insects like moths and beetles.

MOOD MATTERS

Chameleons change color in response to threats or changes in temperature and light. If a chameleon feels calm, it might be green, but if it senses danger, it might turn yellow. Right beneath the chameleon's skin are two cell layers of red and yellow pigments. Underneath are more cell layers that reflect blue and white light. Even deeper is a layer of melanin, the same pigment that gives human skin its variety of shades. Air temperature and sunlight cause changes in hormones, or internal chemicals, to make these cells expand or contract. When the cells get bigger or smaller, different amounts of light pass through the chameleon's transparent skin, and it changes color.

CHEMICAL ARSENAL

Animals also use chemicals for defense or to stun prey.

Skunks spray musk, a strong-smelling chemical made up of thiols and thiol derivatives, from a pair of glands near the base of their tails. Skunks can spray accurately as far as 12 feet. The powerful, unpleasant odor lingers for days and burns if it gets in a victim's eyes. Skunks have very few natural enemies as a result.

Octopuses, squid, and cuttlefish squirt an inky substance into the water when they sense danger. The dark cloud shields them from enemies. The ink is made of mucus and melanin.

Charybdotoxin, apamin, latrotoxin, agatoxin . . . these complicated-sounding terms are names of different types of venom, poisonous chemicals produced by some species of animals. Snakes have hollow fangs that inject venom into their prey. Many fish have sharp spines that deliver venom. Bees, hornets, and wasps have stingers. Spiders usually bite, and scorpions have stingers on their tails.

Different venoms cause different types of damage. Some cause muscle paralysis. Others cause the heart to slow down or stop beating. Still others damage a victim's circulatory system by breaking down the walls of capillaries, causing swelling and bleeding. But the news is not all bad. Some types of venom are used as medicine to treat heart attacks, severe pain, and arthritis and to study nerve impulses.

Activity

CHEMICAL PLANTS Nettles, poison ivy, and other plants have chemical components that can harm animals that touch or eat them. Identify five or six poisonous plants that grow in your area. Find out the names of the harmful chemicals and what symptoms they cause in humans. Are there any remedies for the harm caused by the plants?

KEEPING TABS

Chlorine is an element, a substance that can't be reduced to a simpler substance. In its natural state, it is a toxic greenish gas. Chlorine combines easily with other elements and compounds, and, in combination with other substances, it is found in countless everyday items: table salt, cleaning products, paint, plastics, and paper. Chlorine stays in water supplies, so chemist Ellen Swallow Richards used changes in chlorine levels to monitor changes in water quality. Richards's efforts helped establish the first state water purity standards in the United States.

Boston, Massachusetts, 1888

It was a late night in the Sanitary Laboratory at the Massachusetts Institute of Technology (MIT). Ellen Swallow Richards stared at a map of the state of Massachusetts. She was gathering data on the state's water and sewage systems for the Great Sanitary Survey.

To help her visualize the data, Richards connected areas on the map with similar water chemistry like a dot-to-dot. A form took shape. Richards noticed that the shape of her dot-to-dot images formed lines that followed the shape of the Massachusetts coastline, even if they were inland. This told her that each inland site had levels of contaminants similar to coastal communities. But the most helpful piece of data was the chlorine level. Wind and precipitation brought chlorine compounds from the ocean to other water sources in the state. Unlike other naturally occurring impurities, chlorine remains in water, and it can be reliably tracked. The map showed Richards the current levels of chlorine in water everywhere in the state. This map, later called the Normal Chlorine Map, would become the first assault in the war against water pollution.

Ellen Swallow Richards collecting water samples with a colleague

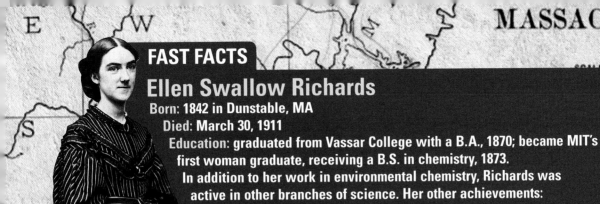

FAST FACTS

Ellen Swallow Richards
Born: 1842 in Dunstable, MA
Died: March 30, 1911
Education: graduated from Vassar College with a B.A., 1870; became MIT's first woman graduate, receiving a B.S. in chemistry, 1873.
In addition to her work in environmental chemistry, Richards was active in other branches of science. Her other achievements:
- Isolated the element vanadium
- Developed a more accurate way to test nickel ore
- Discovered samarskite, an ore

Setting the Stage

The industrial growth of the second half of the 19th century meant bigger cities with crowded living conditions and little in the way of water treatment systems. Richards recognized the dangers of poor water quality and was at the forefront of efforts to ensure a safe water supply. "This is the urban age," she said in 1872, "most of the ills science is called to cure arise from crowded [city] life." In 1885 she noted that if there were 16,600 people living within a 100-mile area, 282 would die from poor environmental conditions. In an area measuring 14 square miles, 415 people would die.

Lasting Contribution

The chlorine levels on the Normal Chlorine Map provided a benchmark for scientists to use in measuring how much of the chlorine pollution in the water occurred naturally and how much was created by humans. The map and its supporting data became a milestone in ecology.

Richards's near-obsession with the environment kept her working on the survey for years. She supervised the testing of 20,000 water samples, making the survey the most comprehensive of its day. Richards and her colleagues also produced the first water purity tables in the world, which showed the types and levels of contaminants.

Lifelong Crusade

Richards received little public credit for her work, but when MIT added a new course on sanitary engineering, Richards was the teacher. Later the Massachusetts Board of Health established the State Water Laboratory—right in Richards's office.

In later years, Richards also studied air quality, metallurgy, and other fields, but she never gave up her pursuit of water ecology. In 1896, MIT published her "Laboratory Notes on Water Analysis." She summed up some of the basics of water study:

The examination of water . . . comprises the determination of three points: first, the amount, if any, of organic matter . . . suspended or dissolved in the water; second, the amount and character of the products of decomposition or organic matter and their relative proportions to each other; and third, the amount of certain mineral substances dissolved in the water.

To make research and data gathering more efficient, Richards identified three classes of water: first, brook, pond, and river water (known as surface water); second, spring and deep-well water; and third, water from shallow wells and from sewage systems.

Richards's work was vital to our understanding of the relationship between water quality and health. And in 1900 she published *Air, Water, and Food*. In the book, she made a profound statement: "Pure water is no longer possible. . . . Safe water is the 20th century's goal."

In the 21st century, the goal is the same.

Activity

QUALITY COUNTS Contact your local water-quality control board and find out how water is tested and purified in your area. Are any chemicals added to disinfect the water? If so, what are they?

VIRTUAL VOYAGE

On a Roll

You've probably had your picture taken hundreds of times. But what turns a roll of film into all those pictures? Besides the artistic talents of the photographer, it's a lot of chemistry.

Someone has taken a picture of you, and suddenly you realize you've been "captured" on a roll of black-and-white film. Say cheese! It's dark inside the camera, and except for a few critical seconds of light, it's going stay that way until you experience a series of chemical changes.

You are photographic film, a strip of plastic 35 millimeters wide. You have another layer of plastic behind you to keep the light coming in through the camera lens from bouncing around and making odd shadows. But what makes you film is your emulsion. That's the gelatin (a super pure form of the stuff that's in gelatin desserts) that's swooshed all over you from head to toe. Scattered throughout the dry gelatin are billions of tiny silver bromide crystals. It's like rolling in beach sand after coming out of the water. Chemically speaking, those crystals are where the action is. When they are exposed to light for even a second, a chemical change occurs in them, which forms the basis of the photograph you will become. Get ready to roll.

Camera, Light, Action

You're stretched across two rollers, then you hear a snap—the camera shutter opens. In a flash, light pours in through the open shutter, reflected off the scene the photographer wants to capture on you. Wherever the light hits those sensitive silver bromide crystals, the exposed crystals undergo a physical change that causes them to be chemically different from the surrounding silver bromide crystals. The crystals not hit by light, the ones that fell in the shadows, do not change. The pattern of the unchanged crystals and the changed crystals will be a black-and-white photograph at the end of the developing process. It's time to move on and make way for the next exposure. Stay out of the light, any more exposure will ruin your image.

Over the next few hours, there are 11 more flashes. You're almost completely transferred to the second roller and are holding a total of 12 images. Almost time for the big move.

24 DISCOVERY CHANNEL SCHOOL

Stages of Development

The camera opens and you pop out, rolled up in a cylinder, protected from light. You enter a room and you're unrolled. It's dark because the silver bromide crystals are still light sensitive. Now you're attached to a cylinder called the film reel and rolled up onto it. SPLASH! You're dropped into a tank full of liquid. It gets darker. A cover has just been placed over the tank. The liquid is a chemical developer called hydroquinone. It reacts with the changed silver bromide crystals to form pure silver grains, which are black. The developer and the crystals that were not exposed to light do not react with each other. The temperature is just right—about 70°F. It has to be, otherwise the developer could damage the image. You start to slosh. Agitation is a also factor in how the chemical process works. The other key factor is timing—too much time and the developer will convert too much of the metallic silver. Speaking of which, it's time to move on.

Next, the stop bath. The developer is poured out and replaced by a mixture of water and acetic acid. By lowering the pH on the film (that is, on you), the mixture stops the action of the developer on the metallic silver and washes the developer off—just like rinsing off soap in the shower.

You've gotten rid of the developer, but you're still carrying around unexposed crystals so you're still light-sensitive. Time to bid them good-bye. The stop bath is emptied and a fixer is added, a chemical such as sodium thiosulfate. More sloshing, and in less than 10 minutes, the unexposed crystals are dissolved. But the black silver remains. You're not light sensitive anymore. You can get out of the dark.

A quick rinse in water, then you're hung up to dry. Check yourself out. drawkcab er'ouY—You're backward. Think about it. That black metallic silver was created where the crystals were exposed to light and the areas that fell in the dark shadows stayed light. That's why you're negative (see picture at left), but it's easy to become positive.

A Positive Finish

Think back to that gelatin emulsion that was swooshed all over you. It also covers photographic printing paper. An enlarging lens is set up between the paper and you. A lamp is turned on, which exposes your image on the paper's silver bromide crystals, so where you are dark, the crystals on the paper stay light and where you are light, the sensitive silver bromide changes. A quick pass through that same series of chemicals, the developer, stop bath, and fixer, and you're totally developed! You're back to your old self, and there you are in the photo too, looking as pretty as a picture.

Activity

LIGHT TOUCH Light plays a role in many chemical reactions. Place one newspaper in an indoor spot where it will get a lot of sunlight, like on a table by a sunny window. Place another newspaper in a dark spot, like the floor of your closet. After five days, compare the two newspapers. What differences do you notice between the two?

HEROES
CRUSADER FOR A

A cure for cancer. It's the goal of countless medical researchers. In their quest, researchers have studied the impact of diet, genetics, and, sometimes, environmental pollutants. But biologist and poet Sandra Steingraber (below) has looked closely at the relationship between pollutants and cancer rates, and her findings have led her to urge researchers to focus much more attention on these contaminants. In her writings and in meetings with countless community groups, she helps the public to become aware of environmental hazards and to look for ways to eliminate the hazards.

Central Illinois, 1980

Soon after her second year of college, Sandra Steingraber is being treated for bladder cancer—extremely rare among young women. There is no obvious cause for her disease. Smoking is linked to bladder cancer, but Steingraber doesn't smoke. Bladder cancer often strikes people in certain occupations, such as mining, printing, and hairdressing—but Steingraber has never held any of these jobs.

Fortunately for her, the cancer was detected relatively early and medical treatment eliminated the disease in her body. Today, after more than two decades, Steingraber remains cancerfree. Steingraber's illness, however, led her to ask some questions. She began to wonder about chemical use in the world around her. Her work toward a doctorate in biology made her aware that perhaps 75,000 different chemicals were regularly used in households and agricultural and industrial processes in the United States—from dry cleaning to car assembly lines to home use of chemicals like paint thinner, cosmetics, and weedkillers.

Moreover, Steingraber learned, these industrial-type chemicals are everywhere. The process of pumping gasoline releases the chemical benzene into the atmosphere. Trace amounts of insecticides remain on crops. The water Steingraber drank when she was a girl, the food she ate, the air she breathed—all were filled with chemicals. Perhaps, she thought, these chemicals had built up in her body and made her more susceptible to disease.

The Chemical History

Steingraber set out to learn as much as she could about the agricultural and industrial chemicals around her. And what she found dismayed her. Increased use of certain types of chemicals has resulted in higher cancer rates. This was of particular interest to Steingraber, who grew up in an area where these chemicals were used. The manufacture of polyvinyl chloride, the material out of which garden hoses and credit cards are made, is connected with higher cancer rates, too. Even chlorinated water has been shown to increase certain types of cancer—including the kind that struck Steingraber.

Just as troubling to Steingraber is the fact that many chemicals in use today have not been examined for potential harm to humans. The pesticide DDT and several other highly poisonous chemicals have been banned. But as Steingraber discovered, many other pesticides that are in use might be just as dangerous. Instead, the U. S. government has set a maximum level of exposure. Being exposed to less than that level is considered "safe."

Also, some harmful chemicals are created by accident. Cancer-causing dioxin can be produced by "newspapers plus plastic wrap plus fire," as Steingraber writes—that is, by burning unsorted garbage. And chemical

SAFER WORLD

 Farmers fly crop dusters to spray pesticides on their fields.

compounds often change when they come into contact with one another. Ground ozone, which can irritate eyes and make breathing difficult on hot days, is produced when exhaust fumes combine with chemicals in substances like dry cleaning fluids or paint.

Steingraber could not trace her own cancer to any particular part of her environment. Nevertheless, the more Steingraber learned, the more alarmed she became. Despite the banning of DDT and a few other substances, industrial chemical use is growing substantially. In 1994, according to one estimate, 2.26 billion pounds of toxic chemicals made their way into the environment. At the same time, the incidence of cancer is growing. Steingraber became convinced that the two were connected.

Hidden Dangers

Some industrial chemicals can become concentrated over time. Insects, their bodies filled with poisonous pesticides, are eaten by birds and fish. These animals cannot eliminate the chemicals from their bodies. The more pesticide-filled insects they eat, the more they become affected by the chemicals. The same holds true for people who eat the birds and fish, or the crops that are sprayed with the pesticides in the first place. While some chemicals leave the body over time, others do not.

Next Steps

Steingraber does not call for immediate banning of all these types of industrial chemicals. She does, however, urge serious study of them by scientists and greater public awareness of the surrounding issues. Despite the depth of the problem, Steingraber is optimistic and believes people can figure out ways to improve matters significantly. Instead of incinerating garbage, we can promote recycling. We can print newspapers with less toxic inks and invent safer processes for dry cleaning. "It is time to start pursuing alternate paths," Steingraber writes in her book, *Living Downstream*. "From the right to know and the duty to inquire flows the obligation to act."

Activity

FARMING FORUM Agricultural pesticide use is controversial. Pesticides may cause harm beyond the targeted pests. On the other hand, they do prevent crop loss. What pests threaten the major crops in your area? What pesticides are used to control them? Are there effective alternatives to the pesticides? Get as much information as you can from your state agriculture department on these issues. Then, divide your class into two teams, and stage a debate on whether or not pesticides should be used on your area's major crop.

SOLVE-IT-YOURSELF MYSTERY

CHEMICAL SOLUTION

Detective Al Keen arrived at the scene of the crime to find Ian Nichols distraught.

"I loved that bike. I'll never be able to save enough to buy a new one," Ian lamented, pointing to the chain-style bike lock that had been cut through, still wrapped through the fence where the bike had been a little while ago.

"This is upsetting," agreed Detective Keen, surveying the area. "Yours is the fifth bike stolen in two weeks. Let's see if our thief left any personal information."

Ian looked at the dangling lock. "Whoever took it left without a trace. I was only gone for ten minutes. I was just dropping off a CD at my friend Bunsen Burns' house on my way home from school. I locked up my bike here to take a shortcut to his house. I left school at about two-thirty, so it must have been around quarter to three."

Detective Keen was already gathering evidence. "Looks like we have some clues to work with," he said. He pulled the lock off the fence and held it up with a gloved hand. "Our forensic chemists can check this for latent fingerprints," he said, dropping it into an evidence bag.

"How?" asked Ian.

"Well, a chemist will heat iodine crystals to make vapors. The vapors will mix with fatty oils in any fingerprints and make the prints visible," answered the detective.

"Hey, look—" Ian pointed to the rosebush coming through the fence. "It looks like someone got scratched on this thorny branch." Sure enough, a few dark droplets were visible on the pale gray sidewalk.

"Good eye, Ian. Our thief snagged some clothing, too," noticed Detective Keen. With tweezers, he pulled blue fibers off the branch. "I'll take these things to the lab and see what our chemists can reveal."

"Will they use iodine vapors again?" Ian asked.

"No, the testing processes will be different. Lab technicians will test and type the blood by looking at how it reacts with different types of blood serums. And they'll examine the fibers under a microscope. If they need more information, they'll analyze the fibers' chemical properties. Before I head to the lab, I'll check with the neighbors to find out if they saw anything suspicious."

By 3 p.m., Crystal Bond was in her overalls, trimming her front hedges. Detective Keen took

28 DISCOVERY CHANNEL SCHOOL

notes as she spoke. "No, I didn't see or hear anything unusual this afternoon. I must have just missed it," she said, stuffing leaves and branches into a garbage bag. "Ouch! I need to be more careful." Crystal revealed her scratched arm.

"Allow me," offered Detective Keen. He wiped the excess blood away with some cotton and put a small bandage on Crystal's arm.

"Thanks," she said. "Sorry I can't be of more help."

"Not a problem," said the detective, putting the cotton in a plastic bag for the lab.

Crystal's next-door neighbor Ag Silvers didn't have any information either. "Well, I'm always out in the afternoon walking the dog. Kryptonite usually barks when strangers come around, but we just saw Ian as he was locking up his bike across the street. When I got back, the bike was gone, so I figured Ian was on his way."

Detective Keen noticed Ag's blue cotton sweatshirt. "We found some blue fibers near where Ian's bike was stolen. Mind if I take a sample from your sweatshirt for the lab?"

Ag raised his eyebrows. "I can assure you, you won't find a match. But if it will prove my innocence, go ahead."

"Many thanks," said the detective, plucking a small fiber off the sweatshirt.

Detective Keen's next stop was Moe LeCule's house, a few blocks away. When he arrived, he found Moe painting the tool shed out back. "Doing a little weatherproofing?" Keen asked.

"Yeah," replied Moe. "I'm really working up a sweat." He mopped his brow with a handkerchief and tossed his fleece vest aside. "I'm sorry to hear about Ian's bike," he said, dipping his brush in a can of metallic paint. "Who would do such a thing?" As Moe flipped the paintbrush, a few splatters of paint landed on Detective Keen's hand. "Sorry," said Moe, handing the detective his handkerchief. Detective Keen wiped the paint off, thanked Moe, and left for the lab.

Two days later, Ian's phone rang. "Hello? Oh, hi, Detective Keen. The Ironman Athletic Shop? Sure, I'll meet you there now."

Ironman Athletics sold used bikes and equipment. "Look at this bike," Detective Keen told Ian. "Do you think it could be yours?"

"It's the same make and model," Ian noted. "But my bike wasn't such a shiny silver. This one looks newer. Besides, I had a different seat and handlebar grips."

Detective Keen grinned. "I got the tip-off from the store owner yesterday about a 'new' used bike. I scraped a small bit of paint off the frame and sent it to the lab. The lab tests are in and the clues point to one of your neighbors."

Who stole Ian's bicycle? Look at the lab results below and see what the forensic chemists discovered.

close up of thorn

Use these clues...

1. The fingerprints on the chain lock matched others found at the scenes of recent bicycle thefts.
2. Chemical analysis of the fibers showed that the material was synthetic—derived from plastics.
3. Both blood samples—Crystal's and the sample from the crime scene—proved to be type-O.
4. About 60 percent of the population can have blood type determined by other bodily fluids, like saliva and perspiration. Swab tests of the perspiration in Moe's handkerchief indicated type-O blood.

Answer on page 32

CHEMISTRY CAPERS

TINY BUBBLES

Joseph Priestly (1733–1804), an amateur English chemist, is considered one of the greatest scientists in history. After all, he identified oxygen, but he is also remembered for another discovery—soda pop. It happened because he was interested in gases and lived near a brewery. This meant Priestly had easy access to carbon dioxide, the gas that forms bubbles in beer and other beverages. With a little experimentation, he was able to combine the carbon dioxide and water. Voilà—the first homemade soda pop.

GREAT ART

Some of the best known paintings in the world aren't painted on canvas. They are frescoes, painted right into the wall or ceiling. One famous fresco was painted by Michelangelo on the ceiling of the Sistine Chapel in Rome, Italy. When artists paint frescoes, they paint directly onto a wet plaster surface. The plaster contains lime, the chemical calcium hydroxide. As the plaster dries, the air reacts with the lime to form calcium carbonate. This causes the paint to crystallize and become part of the plaster.

Mum's the Word

Although they didn't call it chemistry, ancient Egyptians used chemicals to mummify people and favorite animals. In the early efforts, starting about 3200 BC, the goal was to dry out the body. The Egyptians covered a body with natron, the natural mineral sodium carbonate. The natron absorbed water from the body and killed bacteria that might cause decay. Then the body was wrapped in strips of cloth soaked in resin, a liquid derived from plants. By 1500 BC, the process had become more complicated and involved removing organs and filling the abdomen with natron, then soaking the body in more natron for several weeks. The old natron was washed off, and the body was sprinkled with more natron and sweet-smelling spices, and then wrapped up in yards of linen.

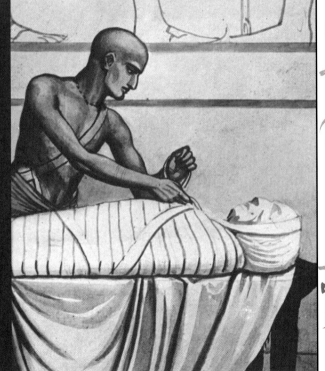

Flower Power

Hydrangeas are sensitive plants—sensitive to acid, that is. If their soil is acidic, the hydrangea flowers will be blue. If the soil is neutral or basic, the flowers will be pink. By the way, you can test your soil's pH and change the color of your hydrangeas by adding certain acids or bases to the soil.

Chemical Expressions

Now that you know a little about chemistry, think about these chemistry-based everyday expressions and compare what they mean in chemistry and in everyday life.

It's dynamite!

Don't over-react!

There's just no chemistry.

I'm neutral on that issue.

Chocolate Lovers

When people fall in love, their brains produce a chemical, phenylethlamine, that increases their heart rate and makes them feel happy. Chocolate contains the very same chemical (though eating it won't produce the same effect). Now you know the link between chocolate and Valentine's Day.

The Red Planet

And think about this: The planet Mars is red because of the element iron. The soil on Mars contains a lot of iron that has combined with oxygen (which was found on Mars long, long ago) to form iron oxides. And it's the oxides that are red.

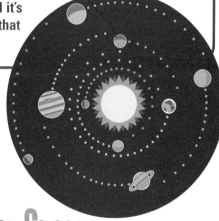

Read My Lips

Have you ever heard of lipsticks that change color based on the wearer's mood? While the lipsticks do change color, it's for a different reason. The pigment, or color, in mood lipstick is acidic. When applied to a non-acidic surface, the lipstick's acidity decreases. That's what makes the color change. The acidity of your lips can change based on what you've eaten, how active you are, and your stress levels. A simple change in mood is not enough.

Storage Space

Consider chemistry when deciding where to store medications, cleaning products, and other household items. The medicine cabinet might not be the best place to keep medicines. Heat and moisture can speed up the deterioration process, which means medicines might lose their effectiveness before the expiration date on the package. For best results, choose a cool, dry, dark place that is out of reach of children. Some cleaning products are not chemically compatible. Chlorine bleach and ammonia produce a toxic gas if they are mixed together, so they should not be stored together.

Taking Stock

YOUR WORLD YOUR TURN

The world is full of all kinds of chemicals—some of them hazardous, others not. You can find many different kinds of chemicals in your own homes. Do a little data gathering: You'll be amazed at what goes into the foods you eat and the products you use around the house every day.

Divide the class into teams and assign each team an area from the list below. With your parent's permission, gather products you find in the assigned area. Using an index card for each product, write down the name of the product (include the brand), the components listed on the label, what the product is used for, and whether there are any warnings or special storage instructions on the label.

Meet with your team and collect the index cards. Compile the information from the cards on a data table listing the name of the product, what it is used for, the chemical components (using a separate line for each), and any warnings or storage instructions you found. Circle the chemicals that are found more than once in your data table. From among the circled chemicals, identify the 10 chemicals that appear most often on the table.

1. Kitchen cabinets—cake mixes, cereals, baking powder, ketchup, mustard, bread, cookies
2. Cleaning supply cabinet—dishwashing detergent, glass cleaner, scouring powder, furniture polish, brass and silver polish
3. Laundry room—detergent, bleach, starch, stain remover
4. Bathroom cabinet—shampoo, soap, lotion, shaving cream, cosmetics
5. Garage—paint, car wax, fertilizer, cleaning supplies

Now use the Internet to find out what each of the ten chemicals is made up of and if any of the chemicals is hazardous. For any hazardous chemicals, see if you can find safer alternatives. For example, a paste of baking soda and water can be used to clean the oven instead of commercial oven cleaner.

As a class, compare the data on hazardous chemicals gathered by all five teams. On butcher paper, compile a data table listing the hazardous chemicals across the top of the page. Down the side, note the product or products in which the chemical is found, any storage instructions or warnings associated with the chemical, and the safer alternatives you found.

Ready for the ultimate challenge? Enter this or any other science project in the Discovery Young Scientist Challenge. Visit *discoveryschool.com/dysc* to find out how.

ANSWER Solve-It-Yourself Mystery, pages 28–29

Moe LeCule was responsible for taking Ian's bike—and the other bikes stolen in the last two weeks. He used his shed as a "chop shop" where he'd repaint the bikes and make simple changes, like adding new handlebars and seats. Then he'd resell them for a profit. The crime lab technicians put both paint samples—from the handkerchief and the bike at Ironman Athletics—in a solvent. From there, they could analyze the elements in the paint pigments. The samples were identical.

The synthetic fibers turned out to be polyester fleece from the vest Moe was wearing that afternoon. Crystal's overalls and Ag's sweater were both made of cotton.

Type-O blood is the most common blood type, so it wasn't too surprising for Crystal's blood type to be the same as that found at the crime scene. The matching of the blood type from the sweaty handkerchief contributed to the strong evidence supporting Moe's guilt.